SpringerBriefs in Applied Sciences and Technology

For further volumes:
http://www.springer.com/series/8884

Seiji Samukawa

Feature Profile Evolution in Plasma Processing Using On-wafer Monitoring System

 Springer

Seiji Samukawa
Tohoku University
Sendai
Japan

ISSN 2191-530X   ISSN 2191-5318 (electronic)
ISBN 978-4-431-54794-5   ISBN 978-4-431-54795-2 (eBook)
DOI 10.1007/978-4-431-54795-2
Springer Tokyo Heidelberg New York Dordrecht London

Library of Congress Control Number: 2013958143

Printed on acid-free paper

Springer is part of Springer Science+Business Media (www.springer.com)

Preface

Recent ultra-large-scale integration (ULSI) production processes have involved the fabrication of sub-22 nm patterns on Si wafers. High-density plasma sources, such as inductively coupled plasma (ICP) and electron-cyclotron resonance (ECR) plasma, are key technologies that enable the development of precise etching processes. However, these technologies cause several types of radiation damage during etching due to charge build-up of positive ions and electrons or radiation from ultraviolet (UV), vacuum ultraviolet (VUV), and X-ray photons. Voltage generated by the charge build-up distorts ion trajectories and leads to the breakage of thin gate oxide films, the stoppage of etching, and etching-rate pattern dependence. Additionally, high-density crystal defects are generated by UV and/or VUV photons radiating from the plasma to the etching surface. These serious problems degrade the electrical characteristics of devices and increase critical dimension losses in the etching process, so they must be overcome to achieve high-quality fabrication of future nanoscale devices. In short, sub-10 nm devices require defect-free and charge-free atomic layer etching processes.

However, the interaction of UV photons and charge build-up on the surface of a given material is not clearly understood because of the difficulty in quantitatively monitoring these problems during plasma processing. In the hopes of making a breakthrough in the prevention of these problems, an on-wafer technique for monitoring the amount of charge build-up and the spectrum of UV photons has been proposed. This technique has been combined with a simulation to establish a relationship between the damage recorded by the on-wafer monitoring technique and the actual damage experienced.

In this book, a good understanding of plasma etching profile is provided with a combination of on-wafer monitoring and etching simulation.

Seiji Samukawa

Contents

Chapter 1
Introduction

Abstract A system to predict etching profiles was developed by combining the results obtained from on-wafer sensors and those obtained from computer simulations. We developed on-wafer UV, on-wafer charge-up, and on-wafer sheath shaped sensors. These sensors could measure plasma process conditions, such as UV irradiation, charge-up voltage in high aspect ratio structures, and ion sheath conditions at the plasma/surface interface on the sample stage. Then, the output of the sensors could be used for computer simulations. The system could predict anomalies in etching profiles around large scale 3D structures that distorted the ion sheath and its trajectory. It could also predict anomalies in etching profiles caused by charge accumulation in high-aspect ratio holes. Moreover, it could predict the distribution of UV-radiation damage in materials.

Keywords Plasma process damage · Charge-up · Ultraviolet photon irradiation · Defect generation · On-wafer monitoring

1.1 Background

Our lives have changed since the invention of integrated circuits (ICs) that are presently incorporated into every major system from automobiles to washing machines and have become the basis of information technology. The IC industry has been expanded by shrinking the feature size of devices in ICs, since miniaturizing devices improves their performance and enables IC chips to contain many more transistors. The feature size of devices has been shrinking and has moved into the nanoscale regime due to Moore's law, which states that the number of transistors on an IC chip will double every 1.4 years.

ICs are fabricated on silicon wafers by repetitive film depositions, lithography, and etching. Plasma is widely used for film deposition and etching and plasma etching processes have particularly contributed to the miniaturization of devices by enabling exact pattern sizes to be transferred to underlayers. Plasma etching has

S. Samukawa, *Feature Profile Evolution in Plasma Processing Using On-wafer Monitoring System*, SpringerBriefs in Applied Sciences and Technology, DOI: 10.1007/978-4-431-54795-2_1, © The Author(s) 2014

also been developed with progress in IC manufacturing to achieve high etch rates, selectivity, uniformity, and control of critical dimensions (CDs) with no damage due to radiation. However, plasma etching to fabricate today's nanoscale devices has become more challenging. The next-generation of nanoscale devices will require the pattern size after etching to be controlled on the atomic scale. Moreover, new innovative technologies have been introduced into these nanoscale devices, such as high-k dielectrics, Cu/low-k interconnects, and novel device structures (FinFETs), which make it more challenging to further develop plasma etching. These process technologies have recently been used to fabricate microelectromechanical systems and nanoelectromechanical systems (MEMS/NEMS). It is necessary to fabricate high aspect ratio structures and three-dimensional structures to produce these devices. This is a new challenge for plasma processes.

1.2 Plasma Etching

Plasma etching is a technology to etch materials with reactive radicals and ions from plasma. Figure 1.1 has a schematic of plasma etching. Since ions can easily be accelerated and collimated, anisotropic and vertical etching is possible. This is one of the major advantages of plasma etching against wet etching where only isotropic etching is possible. Thus, plasma etching is a crucial technology to precisely transfer mask patterns to target materials.

Since parallel-plate reactive-ion etching (RIE) was developed in 1974, many plasma sources have been developed to improve etching characteristics. A crucial limiting feature of parallel-plate RIE is that the ion-bombarding flux and ion-acceleration energy cannot be varied independently. Hence, the sheath voltages at the driven electrode are high for a reasonable ion flux, as well as for a reasonable dissociation of the feedstock gas. This can result in unwanted damage to substrates placed on the driven electrode, or lack of linewidth control. Furthermore, the combination of a low ion flux and high ion energy leads to a relatively narrow window for many process applications. The low process rates resulting from the limited ion flux in RIE often mandate multiwafer or batch processing, with a consequent loss of wafer-to-wafer reproducibility. Higher ion and neutral fluxes are generally required for single-wafer processing in a clustered tool environment, in which a single wafer is moved by a robot through a series of process chambers.

The limitations of parallel-plate RIE and its magnetically enhanced variants have led to the development of a new generation of low-pressure, high-density plasma sources. In addition to high-density and low-pressure, a common feature is that the rf or microwave power is coupled to the plasma across a dielectric window, rather than by direct connection to an electrode in the plasma as in parallel-plate RIE. This noncapacitive electrode power transfer is the key to achieving low voltages across all plasma sheaths at the electrode and wall surfaces. The dc voltages, and hence the ion-acceleration energies, are then typically 20–30 V at all surfaces. The electrode on which the substrate is placed can be independently

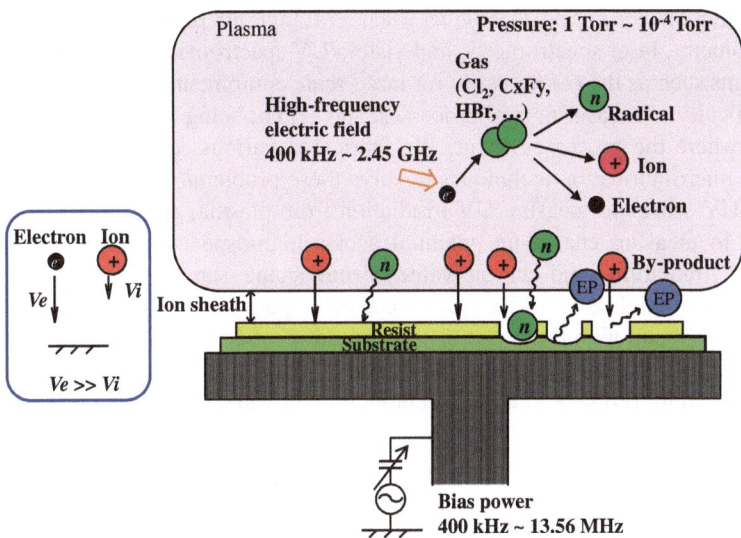

Fig. 1.1 Schematic of plasma etching

driven by a capacitively coupled rf source to control the ion energy. Hence, it is possible to independently control ion/radical fluxes and ion-bombarding energies. Recent ultra-large-scale integration (ULSI) production processes have involved the fabrication of sub-22-nm patterns on Si wafers. High-density plasma sources, such as inductively coupled plasma (ICP) and electron-cyclotron-resonance (ECR) plasma, are key technologies for developing precise etching processes.

However, these technologies create several types of radiation damage caused by the charge build-up of positive ions and electrons [1–4] or radiation from ultraviolet (UV), vacuum ultraviolet (VUV), and x-ray photons [5–12] during etching. Voltages generated by the charge build-up distort ion trajectories and cause thin gate oxide films to break, etching to stop, and the etching rate to depend on patterns. Additionally, high-density crystal defects are generated by UV or VUV photons radiating from the plasma to the etching surface. These serious problems must be overcome in the fabrication of future nanoscale devices as they strongly degrade the electrical characteristics of the devices and increase critical dimension losses in the etching processes. In short, sub-10-nm devices require defect-free and charge-free atomic layer etching processes.

1.3 On-wafer-Monitoring Technique

Process monitoring of wafer positions is needed to attain nanometer-order etching processes. We propose a concept of "on-wafer monitoring", which measures the kinds and energies of active species such as ions, neutrals, radicals, and photons.

Conventional monitoring techniques such as Langmuir probes, quadrupole mass spectrometry, laser spectrometry, and visible/UV spectrometry suffer from various problems such as the (1) necessity for large-scale equipment, (2) disturbed plasma, (3) difficulty of measuring real processes, and (4) not being able to measure on the wafer where the processes occur. We fabricated various sensors using semiconductor microfabrication techniques to solve these problems. We developed an on-wafer UV sensor to measure UV irradiation from plasma, an on-wafer charge-up sensor to measure charge-up potential across high-aspect-ratio structures under plasma irradiation, and an on-wafer sheath shape sensor to measure sheath potential and thickness. Active species and their spatial distribution could be easily monitored in situ with these sensors. We could identify the surface reactions from the measured data. We could also predict process damage distributions and monitor feature profile evolution by combining on-wafer monitoring and computer simulations.

References

1. T. Nozawa, T. Kinoshita, Jpn. J. Appl. Phys. **34**, 2107 (1995)
2. T. Kinoshita, M. Hane, J.P. McVittee, J. Vac. Sci. Technol. **B14**, 560 (1996)
3. H. Ootera, Jpn. J. Appl. Phys. **33**, 6109 (1993)
4. H. Ohtake, S. Samukawa, in *Proceedings of the 17th Dry Process Symposium, Institute of Electrical Engineering of Japan* (Tokyo, 1995), p. 45
5. T. Okamaoto, T. Ide, A. Sasaki, K. Azuma, Y. Nakata, Jpn. J. Appl. Phys. **43**(12), 8002 (2004)
6. K. Yonekura, K. Goto, M. Mastuura, N. Fujiwara, K. Tsujimoto, Jpn. J. Appl. Phys. **44**(5A), 2976 (2005)
7. K.P. Cheung, C.S. Pai, IEEE Electr. Dev. Lett. **16**, 220 (1995)
8. J.-P. Carrere, J.-C. Oberlin, M. Haond, in: *Proceedings of the International Symposium on Plasma Process-Induced Damage* (AVS, Monterey, 2000), p. 164
9. T. Dao, W. Wu, in *Proceedings of the International Symposium on Plasma Process-Induced Damage* (AVS, Monterey, 1996), p. 54
10. M. Joshi, J.P. McVittee, K. Sarawat, in *Proceedings of the International Symposium on Plasma Process-Induced Damage* (AVS, Monterey, 2000), p. 157
11. C. Cismura, J.L. Shohet, J.P. McVittee, in *Proceedings of the International Symposium on Plasma Process-Induced Damage* (AVS, Monterey, 1999), p. 192
12. J.R. Woodworth, M.G. Blain, R.L. Jarecki, T.W. Hamilton, B.P. Aragon, J. Vac. Sci. Technol. A **17**, 3209 (1999)

Chapter 2
On-wafer UV Sensor and Prediction of UV Irradiation Damage

Abstract UV radiation during plasma processing affects the surface of materials. Nevertheless, the interaction of UV photons with surface is not clearly understood because of the difficulty in monitoring photons during plasma processing. For this purpose, we have previously proposed an on-wafer monitoring technique for UV photons. For this study, using the combination of this on-wafer monitoring technique and a neural network, we established a relationship between the data obtained from the on-wafer monitoring technique and UV spectra. Also, we obtained absolute intensities of UV radiation by calibrating arbitrary units of UV intensity with a 126 nm excimer lamp. As a result, UV spectra and their absolute intensities could be predicted with the on-wafer monitoring. Furthermore, we developed a prediction system with the on-wafer monitoring technique to simulate UV-radiation damage in dielectric films during plasma etching. UV-induced damage in SiOC films was predicted in this study. Our prediction results of damage in SiOC films shows that UV spectra and their absolute intensities are the key cause of damage in SiOC films. In addition, UV-radiation damage in SiOC films strongly depends on the geometry of the etching structure. The on-wafer monitoring technique should be useful in understanding the interaction of UV radiation with surface and in optimizing plasma processing by controlling UV radiation.

Keywords On-wafer UV sensor · UV spectrum · Absolute UV intensity · Electron–hole pair · Neural network · Low-k dielectric film

2.1 Introduction

Plasma processes are essential in the fabrication of ultra-large-scale-integrated circuits. There are many activated species in plasma, such as charged particles, radicals, and photons. Etching and deposition processes can be achieved with these activated species. It is important to understand the interaction between plasma and

S. Samukawa, *Feature Profile Evolution in Plasma Processing Using On-wafer Monitoring System*, SpringerBriefs in Applied Sciences and Technology, DOI: 10.1007/978-4-431-54795-2_2, © The Author(s) 2014

surfaces to precisely control plasma etching processes. The interaction of photons with surfaces is not particularly well understood because of the difficulty of monitoring photons during plasma processing. Several studies have reported the effects of photons on surfaces during plasma processing. High energy photons, such as ultraviolet (UV) and vacuum ultraviolet (VUV) photons, generate electron–hole pairs in SiO_2 films, resulting in various types of process damage, such as shifts in the threshold voltage of metal oxide semiconductor transistors [1] and the formation of crystalline defects [2, 3]. Moreover, since these photons can dissociate chemical bonds in sensitive materials, such as low-k dielectric films, ArF photoresist films, and organic materials, they can modify the surfaces of materials and cause process damage in these materials during plasma processing [4–9]. It is necessary to obtain UV spectra and their absolute intensity from plasma to understand what effects photons have on surfaces during plasma processing and predict surface phenomena caused by UV radiation. VUV spectrographs can be used to monitor UV radiation [10, 11] for this purpose. However, VUV spectrographs involve such large and expensive systems that they are difficult to use with plasma tools on commercial production lines. Moreover, UV spectra obtained from spectrographs do not always correspond to UV-radiation incidents that occur on wafers due to different fields of view. We previously proposed an on-wafer monitoring technique that enabled UV photons to be monitored during plasma processing on wafers [12–15] to overcome this issue with VUV spectrographs. We developed newly designed sensors for the technique of on-wafer monitoring on an 8 in. commercial production line for this study, and we established a UV spectrum prediction system, where a UV spectrum and its absolute intensity could be obtained with on-wafer monitoring sensors. We also developed this system to predict low-k dielectric damage during plasma etching [16].

2.2 Experiment

2.2.1 On-wafer Monitoring Technique

Our newly designed on-wafer UV sensors were used in the technique of on-wafer monitoring. The structure of an on-wafer UV sensor is schematically illustrated in Fig. 2.1. On-wafer UV sensors were fabricated on a commercial production line and two embedded poly-Si electrodes in dielectric films were deposited on an Si wafer. The dielectric films on the poly-Si electrodes were 150 nm thick. When an on-wafer UV sensor is irradiated with UV photons with energy that is at a shorter wavelength than the bandgap energy of the dielectric films, UV photons are absorbed in the dielectric films and generate electron–hole pairs. "Plasma-induced current" flows due to the electrons generated by UV radiation by applying dc voltage between the electrodes. We evaluated the UV radiation from plasma as plasma-induced current. Since the bandgap energy depends on dielectric films, we

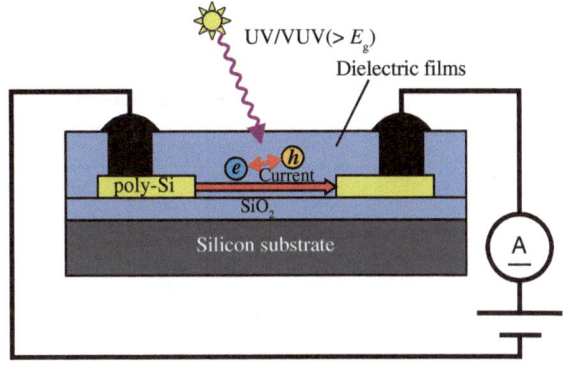

Fig. 2.1 Structure of newly designed on-wafer UV sensor. On-wafer UV sensor has two embedded poly-Si electrodes in dielectric films deposited on Si wafer. Thickness of dielectric films on poly-Si electrodes is 150 nm

can detect different UV wavelength ranges by changing dielectric films on a sensor: SiO_2 for UV photons with energies higher than 8.8 eV (wavelengths shorter than 140 nm) and SiN for UV photons with energies higher than 5 eV (wavelengths shorter than 250 nm). In addition, we used SiN/SiO_2 films for UV photons with energies lower than 5 eV (wavelengths shorter than 250 nm). When UV photons are incident to SiN/SiO_2, UV photons with energies higher than 5 eV (wavelengths shorter than 250 nm) can be absorbed in the SiN layer, resulting in the generation of electron–hole pairs. Since the bandgap energy of SiO_2 is higher than that of SiN, electrons cannot flow in the SiO_2 layer. This means that UV photons with energies higher than 5 eV (wavelengths shorter than 250 nm) do not contribute to plasma-induced current. Other UV photons, on the other hand, with energies lower than 5 eV (wavelengths longer than 250 nm) can penetrate through SiN/SiO_2 layers and be absorbed in the interface between SiO_2 and Si because the energy to generate electron–hole pairs at the interface between SiO_2 and Si is 3.1 eV, which corresponds to 400 nm [17]. Hence, the on-wafer UV sensor with SiN/SiO_2 can detect UV photons with wavelengths longer than 250 nm. A wide range of UV wavelengths can be covered with these three different dielectric films.

Figure 2.2 is a schematic of the measurement setup for the technique of on-wafer monitoring with a plasma chamber. Inductively coupled plasma (ICP) (13.56 MHz) was used to generate high-density plasma of more than 10^{11} cm^{-3}. The on-wafer UV sensors were located on a stage in the plasma chamber where the wafer was usually placed and irradiated with plasma. Lead wires were attached to the electrode pads and connected to an ampere meter and a dc voltage source outside the plasma chamber. We applied 20 V to allow plasma-induced currents to flow between the electrodes. Radio frequency (rf) filters were also used to eliminate rf signals from the plasma during measurements. In addition, a VUV spectrograph was installed at the bottom of the chamber through an 80 mm high and 1 mm diameter pinhole to measure the UV spectrum, and photons were detected at a photomultiplier tube. Electrodes were laterally arranged in the newly designed on-wafer UV sensors, which was different from the previously designed on-wafer UV sensors [12–15]. As the previous on-wafer UV sensors measured plasma-

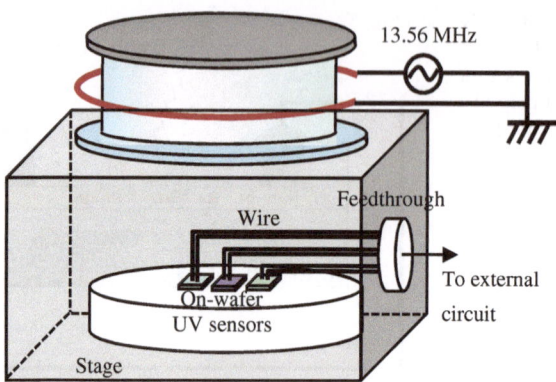

Fig. 2.2 Schematic of measuring setup for on-wafer monitoring technique with plasma chamber. On-wafer UV sensors are located on stage in plasma chamber that wafer is usually placed on and irradiated with plasma

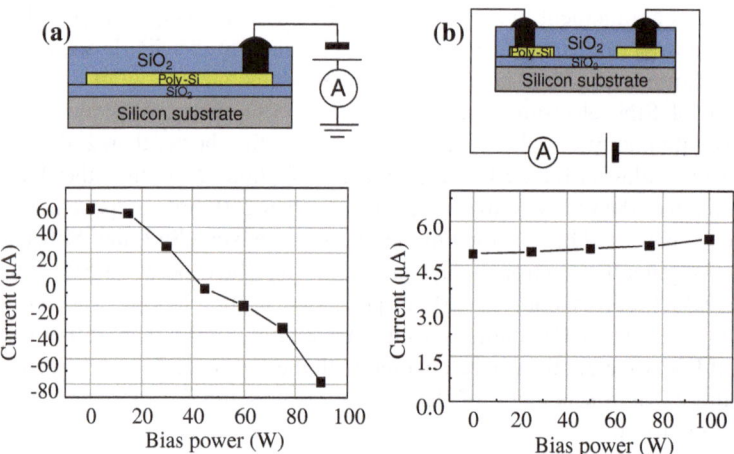

Fig. 2.3 Dependencies of plasma-induced currents on bias power in (**a**) previously designed on-wafer UV sensor and (**b**) newly designed on-wafer UV sensor

induced currents between the surface of the sensor and an embedded poly-Si electrode through dielectric films, the currents depended on the bias power applied to the wafer (Fig. 2.3a). However, currents in the newly designed on-wafer UV sensors were measured between electrodes and did not depend on bias power (Fig. 2.3b). Therefore, our newly designed on-wafer UV sensors could be used even for etching processes with bias power.

2.2.2 Predictive System for UV Spectra

Figure 2.4 shows the predictive system for UV spectra. We used a neural technique of network modeling to relate plasma-induced currents to UV spectra to develop the predictive system. The neural network modeling used in this system is given in Fig. 2.5. Since it has been mathematically proven that a three-layered feed-forward neural network can approximate any functions, the network had a three-layered 3-5-35 neuron model. When plasma-induced currents obtained from three different on-wafer UV sensors were input, the total intensities at intervals of 10 nm in wavelength were output. The network was trained on several data sets of plasma-induced currents and UV intensities measured with a VUV spectrograph. Moreover, the absolute intensities of UV spectra were obtained by calibrating arbitrary units of UV intensity with a 126 nm excimer lamp.

2.2.3 Predictive System for UV-Radiation Damage in Dielectric Films

We improved the predictive system using the technique of on-wafer monitoring to simulate UV-radiation damage in dielectric films during plasma etching processing. Figure 2.6 outlines the predictive flow for UV-radiation damage in dielectric films using this technique. The on-wafer monitoring technique could provide a UV spectrum and its absolute intensity. Furthermore, UV photon trajectories directed to dielectric films were modeled by ray tracing, by taking into consideration etching structures, etching rates, and etching times. We assumed that UV photons were radiated from the edges of ion sheaths in these calculations. UV intensities absorbed in dielectric films were calculated with the absorption coefficient of dielectric films based on the UV intensities incident to dielectric films. UV-radiation damage in dielectric films during plasma etching could be modeled by defining the rates at which defects, or damage, were generated. We simulated etching damage in low-k dielectric films induced by UV radiation, based on the damage predictive system using the technique of on-wafer monitoring. Methyl groups are incorporated to reduce the dielectric constant (k) in low-k dielectric films, such as SiOC films. However, as SiOC films are vulnerable to plasma radiation, such as UV photons, radicals, and ions, SiOC films are severely damaged during plasma etching, particularly on the sidewalls of etching structures. This radiation extracts methyl groups from SiOC films, resulting in an increase in their dielectric constant. We clarified the mechanism responsible for damage to SiOC low-k films during plasma etching [5] and found that UV radiation played an important role in the mechanism, viz., UV breaks Si–C bonds in SiOC films and enhances chemical reactions in SiOC films with radicals and/or moisture. It is

Fig. 2.4 Predictive system for UV spectra

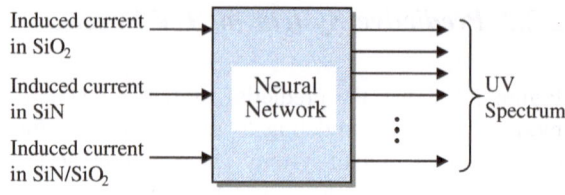

Fig. 2.5 Neural network modeling used in predictive system for UV spectra. Neural network had three-layered 3-5-35 neuron model. When plasma-induced currents obtained from three different on-wafer UV sensors were input, total intensities at intervals of 10 nm in wavelength were output

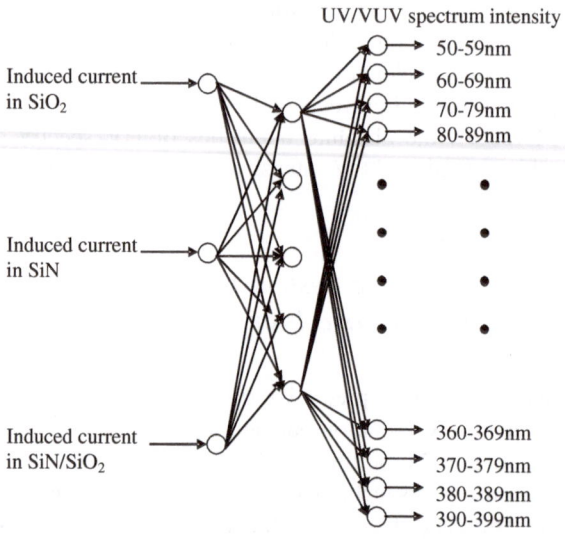

important to be able to predict damage in SiOC films during plasma etching processes to prevent SiOC films from being damaged and optimize etching conditions. We defined UV-induced damage in SiOC films as the breaking of Si–O and Si–C bonds in the predictions since SiOC films consist of Si–O and Si–C bonds and UV photons have enough energy to break these bonds depending on the energy/wavelength of UV photons. Chemical bonds were assumed to have been broken when UV photons had energies higher (shorter wavelengths) than the dissociation energy of bonds: 8.0 eV (150 nm) in Si–O bonds and 4.5 eV (275 nm) in Si–C bonds. The absorption coefficients of these bonds were supposed to be 10^6 cm^{-1} because the absorption coefficients of SiO_2 and SiC dielectric films are almost the same at the value of 10^6 cm^{-1} [18, 19]. We assumed that Si–O and Si–C bonds were broken by UV photons at the same rate if UV photons were above these bond dissociation energies, according to the absorption coefficients of SiO_2 and SiC dielectric films.

Fig. 2.6 Flow for predicting
UV-radiation damage in
dielectric films using on-
wafer monitoring technique

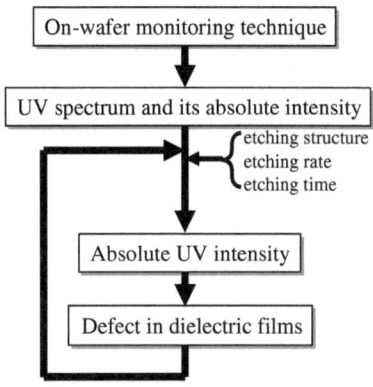

2.3   Results and Discussion

2.3.1   UV Spectrum and Its Absolute Intensity in Plasma

We collected data sets of UV spectra and plasma-induced currents under varying
conditions by changing plasma gases, ICP power, and pressure to train the neural
network used to establish the relationship between plasma-induced currents and UV
spectra. We measured UV spectra with the VUV spectrograph and obtained arbitrary
UV intensities through the measurements. The plasma-induced currents were
measured using three types of on-wafer UV sensors. Figure 2.7 plots the neural
network predictive results for UV intensities in arbitrary units, compared to the
measurements. The plotted data were not used to train the neural network. These
results indicate that the neural network could successfully predict UV intensities.
The absolute intensity of UV spectra was obtained by calibrating arbitrary units of
UV intensity measured with the VUV spectrograph. We used a 126 nm excimer
lamp for this calibration, where the power density of UV light was approximately
5 mW/cm^2 at the lamp window. The lamp was installed in a chamber with the VUV
spectrograph, as seen in Fig. 2.8. The pressure of the chamber was kept at less than
1×10^{-3} mTorr. Photon flux Γ_λ at wavelength λ is described by:

$$\Gamma_\lambda = k\frac{I_\lambda}{tA}, \tag{2.1}$$

where k is the conversion factor, I_λ is the UV intensity (arbitrary units), and t is the
integrated time (0.25 s). Here, A is the irradiated area (0.0079 cm^2). If the conversion factor can be obtained, we can calculate the photon flux of UV light from
the UV intensity. The total power density of photons, P, using Eq. (1.1), can be
expressed as:

$$P = \int E_\lambda \Gamma_\lambda d\lambda = \int E_\lambda \left(k\frac{I_\lambda}{tA}\right)d\lambda, \tag{2.2}$$

Fig. 2.7 Comparison
between UV intensities
measured with VUV
spectrograph and those
predicted with neural network

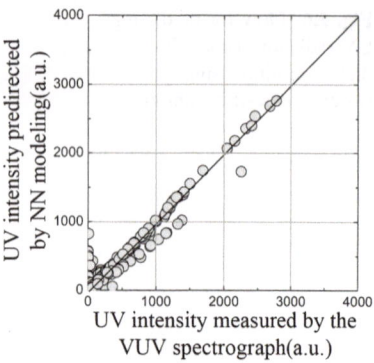

Fig. 2.8 Schematic of
experimental setup for
126 nm excimer lamp and
VUV spectrograph

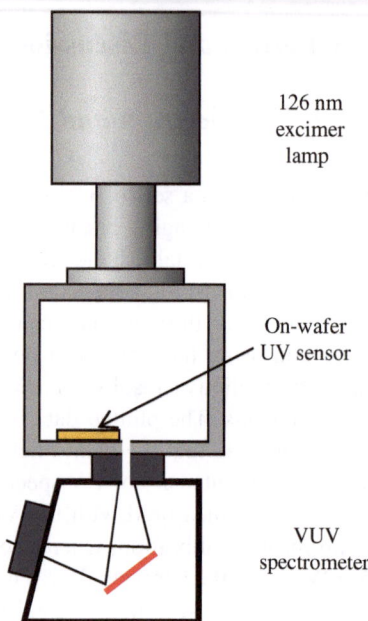

where E_λ is the energy of a photon at a wavelength of λ. The power density of UV
light, on the other hand, decreased as the distance from the lamp window
increased. We measured currents in the sensor, which corresponded to the power
density of UV light, as a function of the distance from the lamp window by
irradiating an on-wafer UV sensor with UV light from the lamp, as shown in
Fig. 2.9. We could acquire the following empirical equation from these results for
the dependence of power density of UV light on distance:

$$P = P_0 10^{-\alpha L}, \tag{2.3}$$

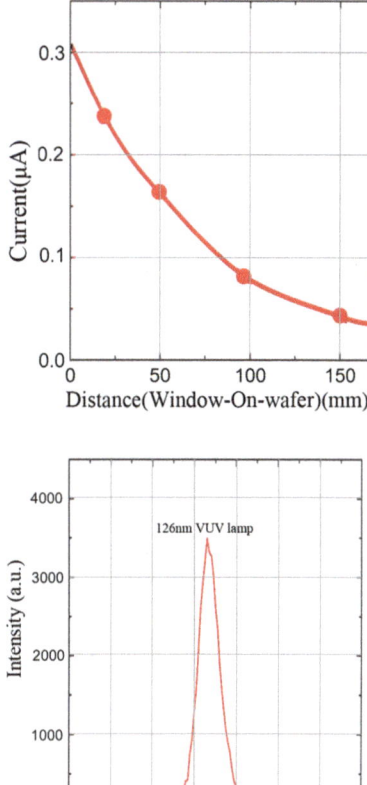

Fig. 2.9 Currents in on-wafer UV sensor (SiO_2) under lamp as function of distance from lamp window

Fig. 2.10 Spectrum of lamp measured with VUV spectrograph

where P_0 is the power density of UV light at the lamp window (5 mW/cm^2), α is a constant (0.006), and L is the distance from the lamp window. The spectrum of the lamp measured with the VUV spectrograph is shown in Fig. 2.10, where a peak can be observed in the wavelength of 126 nm. We obtained $k \sim 2 \times 10^9$ by integrating UV intensities in the spectrum of the lamp and equating Eqs. (2.2) and (2.3) since the detector was located 230 mm from the lamp window. Using this conversion factor enabled us to calculate the absolute intensities of UV spectra.

Figure 2.11 has examples of UV spectra directly measured with the VUV spectrograph ("measurements") and those obtained using the on-wafer monitoring technique and calibration ("prediction") in Ar, CF_3I, and C_4F_8 plasmas under conditions of 1000 W of ICP power, 20 SCCM (SCCM denotes cubic centimeter per minute at STP) of mass flow, and 5 mTorr of pressure. The UV spectra with the technique of on-wafer monitoring agreed well with those measured with the VUV spectrograph. In addition, the comparison of absolute intensities of UV photons in this study and those in the previous study provided reasonably good

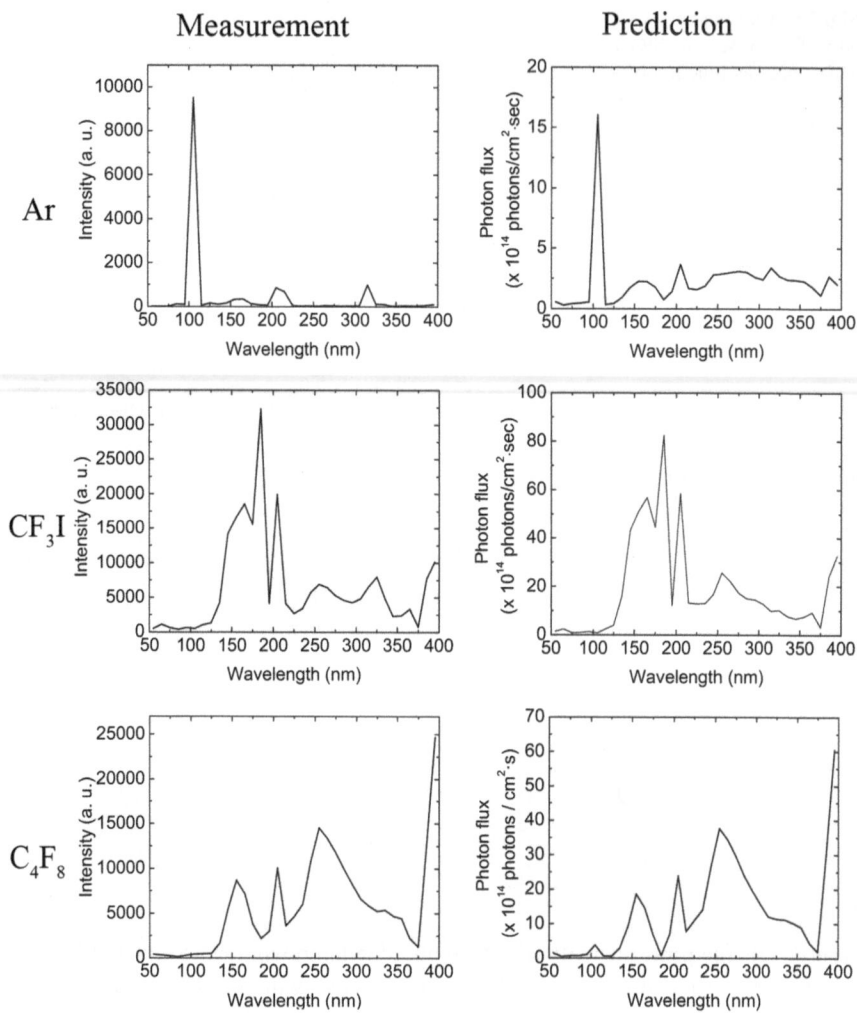

Fig. 2.11 Examples of UV spectra directly measured with VUV spectrograph (measurements) and those obtained with on-wafer monitoring technique (prediction) in Ar, CF_3I, and C_4F_8 plasmas under conditions of ICP power of 1000 W, mass flow of 20 SCCM, and pressure of 5 mTorr

agreement. Woodworth [11] reported that the absolute intensity of UV photons in a wavelength range from 70 to 140 nm in C_4F_8 plasma where $n_e = 3.0 \times 10^{11}$ cm^{-3} and $T_e = 4$ eV was 3.0×10^{15} cm^{-2} s^{-1} based on measurements with a VUV spectrograph. However, the absolute intensity of UV photons in the same range was about 1×10^{15} cm^{-2} s^{-1} under similar plasma conditions (C_4F_8 plasma, $n_e \sim 3 \times 10^{11}$ cm^{-3}, and $T_e \sim 4$ eV). This means that the technique of on-wafer monitoring could successfully provide an UV spectrum and its absolute intensity during plasma processing.

Fig. 2.12 UV spectrum and its absolute intensities in CF$_4$ plasma obtained with on-wafer monitoring technique

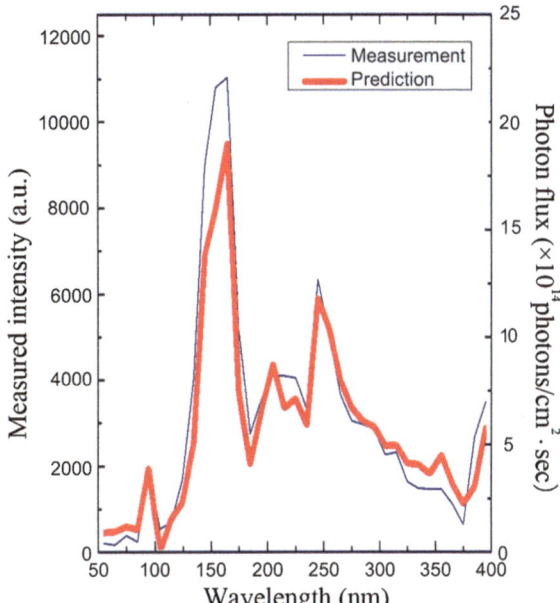

2.3.2 UV-Radiation Damage in Low-k Dielectric Films

We used CF$_4$ plasma to predict damage from radiation. The UV spectrum and its absolute intensities in CF$_4$ plasma were obtained using the technique of on-wafer monitoring, as seen in Fig. 2.12. The etching structure of SiOC films with hard masks in the trench structure was modeled, as can be shown from Fig. 2.13. We assumed that the etching rate was 60 nm/min, the etching time was 100 s (just etching), and that the hard mask perfectly absorbed UV photons, viz., there were no transmissions or reflections from the hard mask during plasma etching. The results for predicting damage in Si–O and Si–C bonds at etching depths of 25, 50, and 100 nm are given in Fig. 2.14. These results give us several insights into UV-radiation damage in SiOC films during plasma etching. First, no great damage can be observed at the bottom of the trench structure, compared to its sidewalls. This is because the damaged layers at the bottom were removed during plasma etching. Second, the damaged layer in Si–C bonds was much larger than that in Si–O bonds. This means that most of the damage during the etching of SiOC films with CF$_4$ plasma was not induced in Si–O bonds, but in Si–C bonds because the Si–C bonds were sensitive to a wider range of UV radiation than Si–O bonds. In addition, the UV spectra in the plasmas were the keys to the cause of damage in SiOC films. There is a smaller number of photons in wavelength ranges of less than 150 nm according to the UV spectrum in CF$_4$ plasma, compared to ranges of less than 275 nm, which can also explain the difference in damage between Si–O and Si–C bonds. Finally, it was apparent that the damage layer increased as

Fig. 2.13 Model of etching structure of SiOC films with hard masks in trench. Etching rate was 60 nm/min, etching time was 100 s (just etch), and hard mask perfectly absorbed UV photons, viz., there were no transmissions or reflections in hard mask during plasma etching

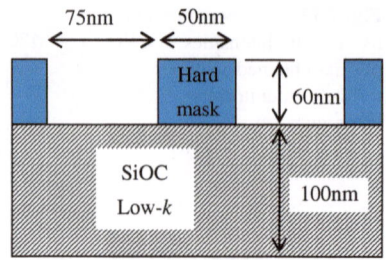

Fig. 2.14 Results from predicting damage in Si–O and Si–C bonds at etching depths of 25, 50, and 100 nm

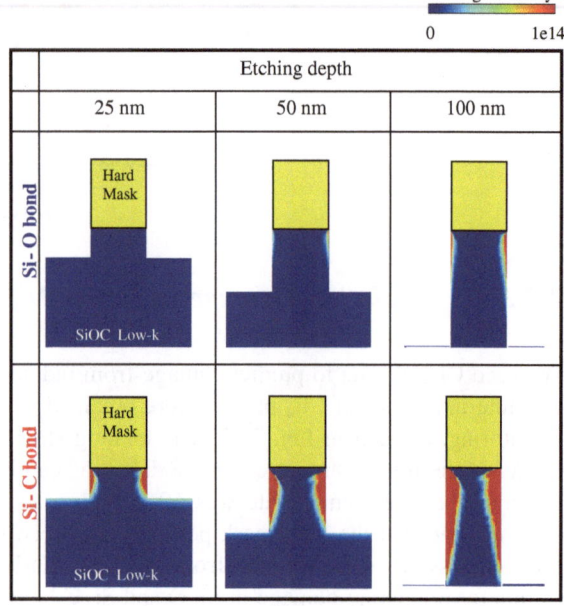

etching progressed but damage was not found underneath the hard mask. This indicated that UV radiation was shaded by the hard mask. In other words, UV-radiation damage in SiOC films strongly depended on the geometry of the etching structure. The etching damage profile in SiOC films during CF_4 plasma was experimentally investigated by Iba [20]. The etching damage profile was close to the damaged layer in Si–C bonds. This means that UV photons mainly affected the side-wall surface of SiOC films during plasma etching and induced damage by breaking the bonds and enhanced chemical reactions with radicals and/or moisture. If damage was only caused by radicals, it would have been observed even underneath the hard mask because radicals move isotropically and collide with other particles. It was difficult to induce deep damage into sidewalls near the hard mask by ion bombardment because ions were accelerated by rf bias applied to the wafers. Therefore, the results from prediction clarified that UV spectra and their absolute intensities were important in the formation of damage since damage in SiOC films strongly depends on UV spectra and their absolute intensities.

2.4 Conclusions

UV spectra and their absolute intensities during plasma processing were predicted with our technique of on-wafer monitoring. We established a neural network to relate plasma-induced currents obtained with this on-wafer monitoring and UV intensities measured with a VUV spectrograph. We also calculated the absolute intensities of UV photons by calibrating arbitrary units of UV intensity with a 126 nm excimer lamp. UV spectra could be successfully predicted, and their absolute intensities predicted with our on-wafer monitoring were consistent with those measured with a VUV spectrograph in a previous report. Moreover, we improved the predictive system with on-wafer monitoring to simulate damage in SiOC low-k films during CF_4 plasma etching. The predicted damage profiles of SiOC films were similar to the experimentally obtained damage profiles. We found from these results of prediction that UV radiation damages the Si–C bonds of SiOC films during plasma etching. In addition, our results indicated that UV-radiation damage in SiOC films strongly depended on the geometry of the etching structures. On-wafer monitoring should be useful in understanding the interaction of UV radiation with surfaces and in optimizing plasma processing by controlling the effects of UV radiation.

References

1. T. Yunogami, T. Mizutani, K. Suzuki, S. Nishimatsu, Jpn. J. Appl. Phys. Part 1 **28**, 2172 (1989)
2. T. Tatsumi, S. Fukuda, S. Kadomura, Jpn. J. Appl. Phys. Part 1 **32**, 6114 (1993)
3. T. Tatsumi, S. Fukuda, S. Kadomura, Jpn. J. Appl. Phys. Part 1 **33**, 2175 (1994)
4. D. Nest, D.B. Graves, S. Engelmann, R.L. Bruce, F. Weilnboeck, G.S. Oehrlein, C. Andes, E.A. Hudson, Appl. Phys. Lett. **92**, 153113 (2008)
5. B. Jinnai, T. Nozawa, S. Samukawa, J. Vac. Sci. Technol. B **26**, 1926 (2008)
6. B. Jinnai, K. Koyama, K. Kato, A. Yasuda, H. Momose, S. Samukawa, J. Appl. Phys. **105**, 053309 (2009)
7. E. Soda, N. Oda, S. Ito, S. Kondo, S. Saito, S. Samukawa, J. Vac. Sci. Technol. B **27**, 649 (2009)
8. S. Uchida, S. Takashima, M. Hori, M. Fukasawa, K. Ohshima, K. Nagahata, T. Tatsumi, J. Appl. Phys. **103**, 073303 (2008)
9. S. Samukawa, Y. Ishikawa, K. Okumura, Y. Sato, K. Tohji, T. Ishida, J. Phys. D Appl. Phys. **41**, 024006 (2008)
10. J.R. Woodworth, M.G. Blain, R.L. Jarecki, T.W. Hamilton, B.P. Aragon, J. Vac. Sci. Technol. A **17**, 3209 (1999)

11. J.R. Woodworth, M.E. Riley, V.A. Arnatucci, T.W. Hamilton, B.P. Aragon, J. Vac. Sci. Technol. A **19**, 45 (2001)
12. S. Samukawa, Y. Ishikawa, S. Kumagai, N. Okigawa, Jpn. J. Appl. Phys. Part 2 **40**, L1346 (2001)
13. M. Okigawa, Y. Ishikawa, S. Samukawa, J. Vac. Sci. Technol. B **21**, 2448 (2003)
14. Y. Ishikawa, Y. Katoh, M. Okigawa, S. Samukawa, J. Vac. Sci. Technol. A **23**, 1509 (2005)
15. M. Okigawa, Y. Ishikawa, S. Samukawa, J. Vac. Sci. Technol. B **23**, 173 (2005)
16. B. Jinnai, S. Fukuda, H. Ohtake, S. Samukawa, J. Appl. Phys. **107**, 043302 (2010)
17. R. Williams, Phys. Rev. **140**, A569 (1965)
18. H.R. Philipp, D.P. Edward, *Handbook of Optical Constants of Solids* (Academic, Burlington, 1997), p. 719
19. W.J. Choyke, E.D. Palik, D.P. Edward, *Handbook of Optical Constants of Solids* (Academic, Burlington, 1997), p. 587
20. Y. Iba, S. Ozaki, M. Sasaki, Y. Kobayashi, T. Kirimura, Y. Nakata, Mechanism of porous low-k film damage induced by plasma etching radicals. Microelectron. Eng. **87**, 451 (2010)

Chapter 3
Prediction of Abnormal Etching Profiles in High-Aspect-Ratio Via/Hole Etching Using On-wafer Monitoring System

Abstract For the prediction of abnormal etching profiles, an ion trajectory prediction system has recently been developed. In this system, sheath modeling was combined with the on-wafer monitoring technique for accurate prediction. This system revealed that sidewall conductivity strongly affects the charge accumulation and ion trajectory in high-aspect-ratio holes. It was also found that the accumulated charge in adjacent holes is one of the reasons for the generation of twisting profiles according to analysis using the system. We presume that the prediction system is an effective tool for developing nanoscale fabrication.

Keywords On-wafer charge-up sensor · High-aspect-ratio holes · Bowing · Etch stop · Twisting · Sidewall conductivity

3.1 Introduction

Ultra-large-scale integrated circuit (ULSI) devices have many high-aspect ratio structures, such as shallow trench isolation (STI) structures, cylinder capacitors, and through-silicon vias (TSVs). These structures are fabricated with plasma etching processes. However, the generation of abnormal profiles, such as bowing, etch stops, and twisting, has been reported in high-aspect-ratio hole etching (Fig. 3.1). Twisting, in particular, is one of the severest problems in nanoscale device fabrication. Some researchers have pointed out that abnormal profiles such as those of twisting are caused by the distortion of ion trajectories [1, 2]. This distortion of ion trajectories is considered to result from the bias of charge accumulation in holes. There is an ion sheath in front of the wafer in plasma etching due to the energy difference between ions and electrons. Ions are accelerated into holes by the ion sheath, while electrons cannot enter the holes due to their isotropic velocity distribution. This is the so-called "electron shading effect" [3]. The bottom of contact holes is positively charged, which significantly affects ion trajectories. We have to observe and precisely control charge accumulation on

S. Samukawa, *Feature Profile Evolution in Plasma Processing Using On-wafer Monitoring System*, SpringerBriefs in Applied Sciences and Technology, DOI: 10.1007/978-4-431-54795-2_3, © The Author(s) 2014

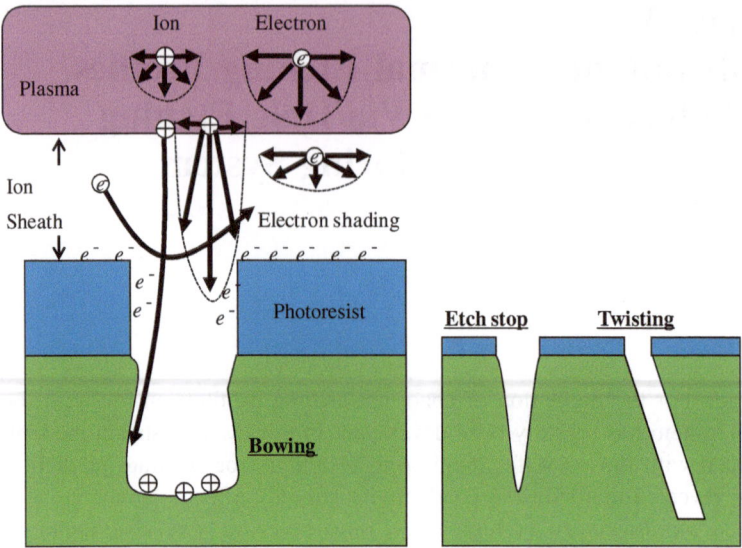

Fig. 3.1 Bowing, stopped etching, and twisting in high-aspect ratio holes

wafer surfaces to avoid ion trajectory distortion and twisting profiles. Some researchers have investigated ion trajectory predictions [4–12]. The plasma structure and sheath area have been simulated to determine potential in most of their investigations. However, there have been several problems with such predictions. One of the most serious has been the difference between actual and simulation values. Outside conditions, such as pressure, source and bias powers, and gas species, have been used for the boundary conditions [4, 5] in some simulations. However, since the actual surface conditions may change for various reasons, the results from simulation do not correspond to the actual ones. The ion/electron flux and energy have been given by the authors [6–12] in other simulations. The results in these cases do not match the actual results. We have already developed and reported a technique of on-wafer monitoring to directly obtain information on the wafer surface. We found out what the electron temperature, electron density, ion energy, and sidewall resistance of holes [13–17] were. An electron temperature and density sensor revealed a lower electron density and a higher electron temperature at the bottom of contact holes due to the electron shading effect than those in bulk plasma [13]. The charge-up sensor revealed that the electron shading effect could be clearly observed as the potential difference between the wafer surface and the bottom of contact holes [14, 15]. It was also revealed that a sidewall-deposited fluorocarbon film even in high-aspect-ratio contact holes has high electric conductivity, which may mitigate electric charge accumulating at the bottom of contact holes during SiO_2 etching processes [16, 17]. We developed a new ion-trajectory prediction system in high-aspect-ratio holes in this study by combining the technique of on-wafer monitoring and sheath modeling to explain and predict twisting phenomena. The accuracy of simulation was increased by using the data from on-wafer monitoring sensors as the boundary conditions. We tried to predict the potential difference between the surface and the bottoms of holes in etching them with an aspect ratio of 15 and confirmed that the simulation data corresponded to the measured data. The distortion of ion trajectories in the generation of twisting profiles was predicted after the validity of the simulations was tested and proved [18].

3.2 Experimental and Simulation Models

We considered the etching of holes with a diameter of 100 nm in this study. Figure 3.2 shows the simulation sequence where on-wafer sensors provided the electron temperature, electron density, surface potential, and sidewall resistance. The ion/electron motion and field potential near the wafer surface were calculated self-consistently by using these data as boundary conditions. The ion trajectory could easily be calculated when the field potential was known. Figure 3.3 shows the on-wafer sensors we used in this study. An on-wafer monitor was placed at the bottom of the plasma reactor during the experiment. The signals were passed outside the chamber with a lead wire. The noise from the plasma was reduced using a voltage/current measurement system with an RF filter. Figure 3.3a shows the on-wafer probe we developed to measure the electron temperature and electron density [13]. The on-wafer probe for electron energy was a stacked structure of Al_2O_3 (280 nm thick) and aluminum films. The Al_2O_3 film was fabricated by anodically oxidizing aluminum. The diameter of the patterned holes was 500 nm. There were 4,800,000 holes in the monitoring device. The exposed area was 0.0942 cm^2. Figure 3.3b shows the charge-up sensor [15]. Two polycrystalline silicon electrodes were separated by a 1.2-mm-thick SiO_2 film. The bottom poly-Si electrode under the SiO_2 layer was 300 nm thick. The diameter of the patterned contact holes was 100 nm and there were 150,000,000 contact holes in the monitoring device. The surface and hole-bottom potentials were directly measured with the sensor. In addition, the sidewall resistance was also measured by measuring the resistance between the top and bottom electrodes in this sensor.

Figure 3.4a shows the simulation model used in this work, which corresponds to the structure of the charge-up sensor. We considered etching a hole with a diameter of 100 nm. We monitored ion and electron motions under a field potential in these calculations. The governing equations are motion equations of ions and electrons, and the Poisson equation:

$$M_e \frac{d^2 r}{dt^2} = -eE, \qquad (3.1)$$

$$M_p \frac{d^2 r}{dt^2} = eE, \text{ and} \qquad (3.2)$$

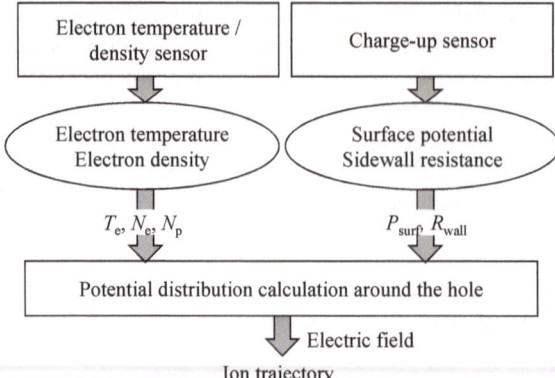

Fig. 3.2 Concept underlying system for ion-trajectory predictions in this study. Accurate predictions of ion trajectories could be achieved using measured values around holes from on-wafer sensors (T_e: electron temperature, N_e: electron density, N_p: plasma density, P_{surf}: surface potential, and R_{wall}: sidewall resistance)

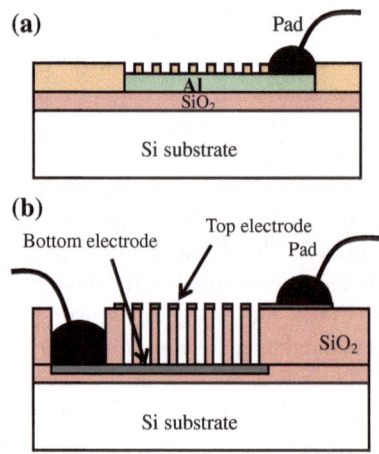

Fig. 3.3 **a** Electron temperature/density monitoring sensor and **b** charge-up sensor

$$\triangle \phi = -\frac{e(n_p - n_e)}{\varepsilon_0 \varepsilon_r}, \tag{3.3}$$

where M_e is the electron mass, M_p is the ion mass, and r is the position vector. Here, t is the time, e is the elementary charge, and E is the electric field. The φ is the field potential, n_p is the ion density, and n_e is the electron density. The ε_0 is the permittivity of free space and ε_r is the relative permittivity.

The ions and electrons were emitted 1.5 mm apart from the wafer and entered the wafer. The surface charge increased as ions were injected or decreased electrons were injected. The motions of ions and electrons were affected by the field

Fig. 3.4 **a** Simulation model in this study. **b** Charge accumulation model at sidewall of holes taking into consideration sidewall conductivity

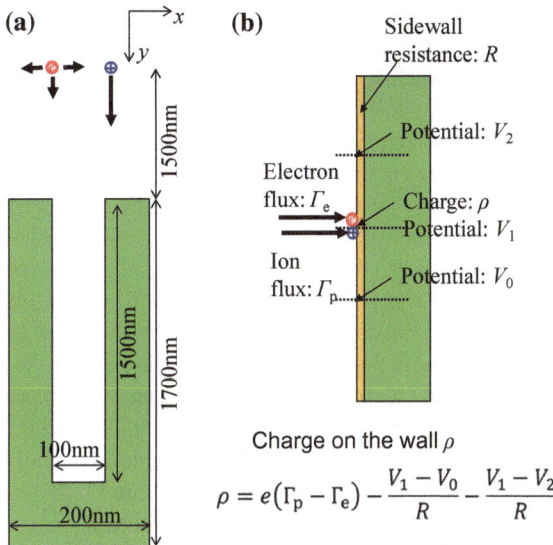

potential generated by the accumulated charge. The calculations were repeated until the field potential demonstrated no change. The ion/electron motions and field potential were solved self-consistently according to this sequence. The surface charge accumulation has been treated in Fig. 3.4b. The accumulated charge decreased with the sidewall current, followed by sidewall resistance. Then, the equation is:

$$\rho = e(\Gamma_p - \Gamma_e) - \frac{V_1 - V_0}{R} - \frac{V_1 - V_2}{R}, \tag{3.4}$$

where ρ is the accumulated charge, Γ_p is the ion flux, and Γ_e is the electron flux. Here, V_1 is the potential during the injection of electrons and ions and V_0 and V_2 are the potentials of adjacent cells.

3.3   Results and Discussion

3.3.1 Ion-Trajectory Prediction

Figure 3.5 shows the calculated potential distribution around the SiO_2 hole with an aspect ratio of 15 (depth: 1.5 mm). The actual values measured with the on-wafer sensors were used (surface potential: –42 V, electron temperature: 4.3 eV, and electron density: 4×10^9 cm^{-3}) in the simulations. The bottom of the hole was positively charged, which was due to the electron shading effect. Figure 3.6 shows the calculated potential distribution and ion trajectory as a function of sidewall resistance when the effect of sidewall resistance was observed. In all cases, positive charges due to electron shading were observed. However, the charge-up potential, which is defined as the potential difference between the surface and the bottom of the hole, drastically decreased with decreasing sidewall resistance. This indicates that the sidewall current reduced the positive charge at the bottom of the hole. In addition, the ion trajectories were distorted by the varying field potentials. Thus, we have to carefully determine field potential to precisely predict the ion trajectories.

We compared the simulated charge-up potentials with those from the experiments to test and prove the validity of the system. Figure 3.7a shows the calculated and measured charge-up potentials (the potential difference between the top and bottom of the hole) as a function of sidewall resistance. The charge-up potential drastically decreased with decreasing sidewall resistance in both cases. The simulated charge-up potential was almost the same value as the one that was measured. This indicates that the experimental data can be predicted with this system, proving the validity of this method. Figure 3.7b shows the mechanism for charge reduction at low sidewall resistance. Electrons accumulate at the surface due to the energy difference between electrons and ions regardless of sidewall resistance. Electrons cannot enter the hole at that time because of electron shading. Charge separation is not solved at a high sidewall resistance because there is no transport of charges. However, electrons can move from the wafer surface to the bottom of the hole at low sidewall resistance, which causes reduced positive charge at the bottom of the hole. As a result, we found that sidewall resistance played an important role in determining field potential and ion trajectories.

3.3.2 Twisting Predictions

Some researchers have pointed out that twisting profiles are caused by the bias of charge accumulation due to malformed resists and deposited films [1, 2, 19–22]. However, the precise mechanisms for the generation of twisting profiles remain unknown. Figure 3.8a has an example of twisting profiles observed in our laboratory [inductively coupled plasma (ICP) etcher, $Ar:C_4F_8 = 9:1$, total flux: 30 sccm, pressure: 30 mTorr, ICP power: 1 kW (13.56 MHz), and bias power: 100 W (1 MHz)]. This was a dense hole pattern where the sizes of the holes were 100 nm and their spaces were 100 nm. However, the actual sizes of the holes and spaces changed as the resists deformed. We can see that the hole profiles are distorted from the scanning electron microscopy (SEM) image. We can also observe that the distortion in holes increased as the pattern spaces decreased. Figure 3.8b shows the relationship between pattern spaces and the taper angles of holes in Fig. 3.8a. Pattern space was defined as the pattern space on the SiO_2 surface because the resist was deformed. The taper angle was defined as the taper angle on the right and at the bottom of a hole located on the left of the measured space. This was because the SEM image was clear. The taper angle decreased with

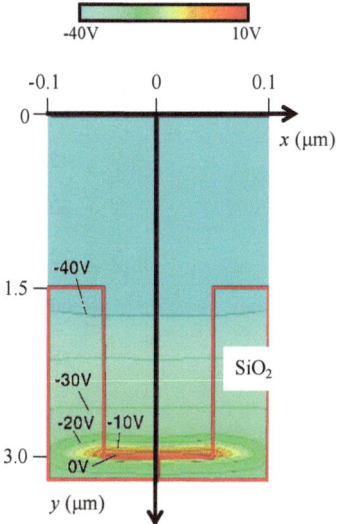

Fig. 3.5 Calculated potential distribution around SiO_2 hole (sidewall resistance: 3×10^{15} Ω, surface potential: -42 V, electron temperature: 4.3 eV, and electron density: 4×10^9 cm^{-3})

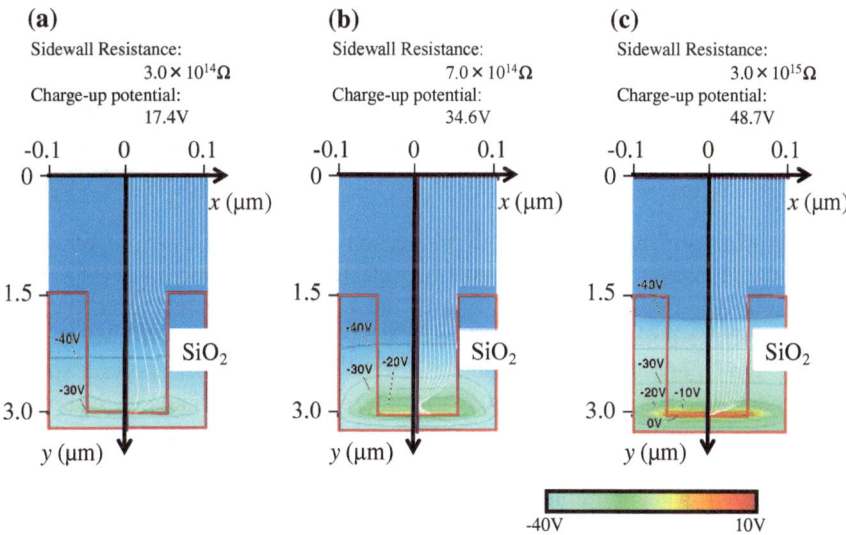

Fig. 3.6 Calculated potential distribution and ion trajectory (*white line*) at sidewall resistance of **a** 3×10^{14}, **b** 7×10^{14}, and **c** 3×10^{15} Ω (surface potential: -42 V, electron temperature: 4.3 eV, and electron density: 4×10^9 cm^{-3})

decreasing pattern space. Although charge accumulation was thought to be one of the reasons for the twisting profiles to generate, the effect of pattern space has not yet been discussed. Then, we attempted to predict the ion trajectory for generating

Fig. 3.7 a Calculated and
measured charge-up voltages
as function of sidewall
resistance. **b** Mechanism for
reduced charge-up by
decreasing sidewall
resistance

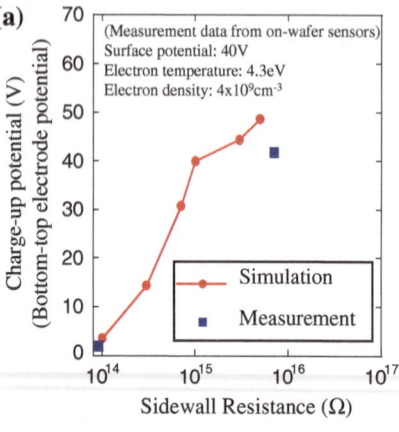

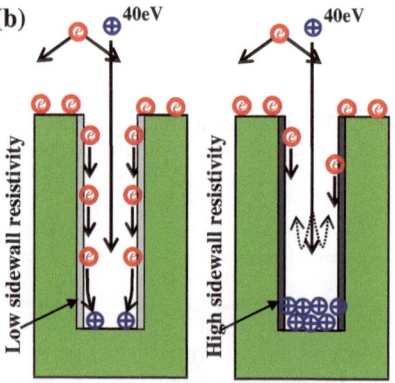

twisting profiles using the scheme to predict ion trajectories we developed, which
could predict charge accumulation and ion trajectories in isolated holes (Fig. 3.7a).
The results from on-wafer monitoring in the plasma shown in Fig. 3.8a are as
follows: surface potential: −35 V, electron temperature: 4.0 eV, electron density:
5×10^9 cm^{-3}, and sidewall resistance: 3×10^{15} Ω. Figure 3.9 shows a simulation model that has two holes. Actually, the effect of accumulated charges in
adjacent holes on the ion trajectory is very complicated because there are many
holes and resist deformation cannot be predicted. The ion trajectory in this study
was predicted using minimum units, i.e., two holes, to observe what effect the
accumulated charges in the most adjacent hole had on the ion trajectory when the
pattern space decreased due to the resist deforming.

Figure 3.10 shows the calculated potential distribution and ion trajectory as a
function of the spaces between the holes at an aspect ratio of 20. The actual values
measured with on-wafer sensors were used in the simulation (surface potential: −
35 V, electron temperature: 4.0 eV, electron density: 5×10^9 cm^{-3}, and sidewall
resistance: 3×10^{15} Ω). The ion trajectory in small spaces was distorted more
than that in large spaces. The ions in small spaces (20 or 40 nm) accelerated to the

Fig. 3.8 a SEM image of contact holes. **b** Dependence of taper angle on pattern space

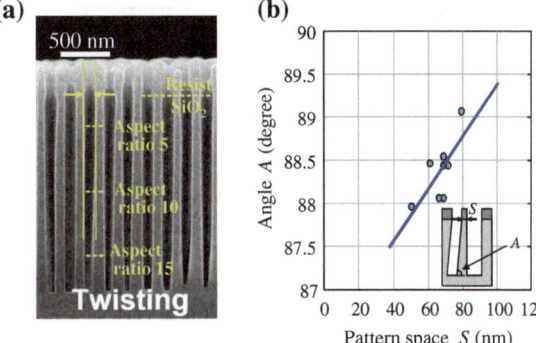

Fig. 3.9 Simulation model for evaluating generation of twisting profiles

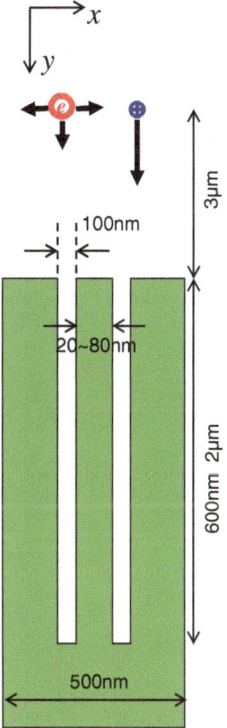

opposite side of adjacent holes. The incident angle of ions as a function of pattern space is presented in Fig. 3.11 to clarify the difference between the results. This figure clearly indicates that the ion trajectory was distorted in a small pattern space. The simulated taper angle in a pattern space of 40 nm was almost the same as the experimental taper angle (Fig. 3.8b, $\sim 88°$).

Figure 3.12 shows the calculated potential distribution and ion trajectory as a function of the aspect ratio with a pattern space of 20 nm. The actual values

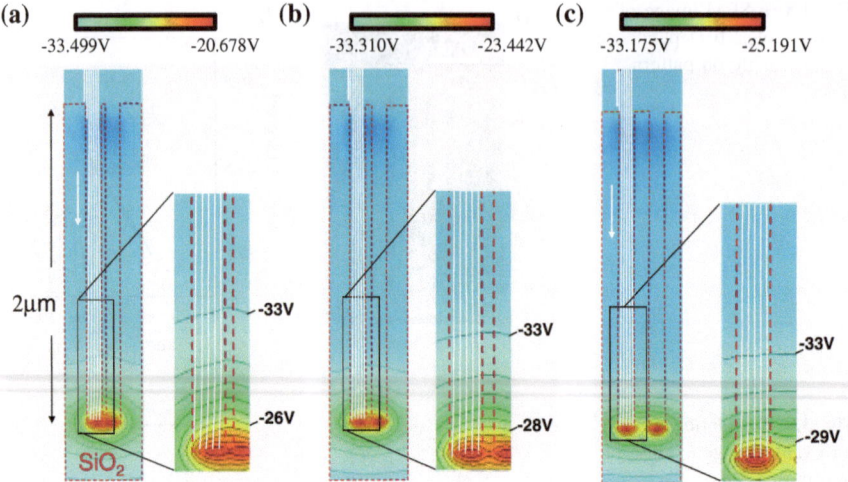

Fig. 3.10 Calculated potential distribution and ion trajectory (*white line*) for pattern spaces of (**a**) 20, (**b**) 40, and (**c**) 100 nm at aspect ratio of 20. As space decreases, ion trajectory is distorted further (surface potential: 35 V, electron temperature: 4.0 eV, electron density: 5×10^9 cm^{-3}, and sidewall resistance: 3×10^{15} Ω)

Fig. 3.11 Incident angle of ions to bottom of hole under conditions in Fig. 3.10

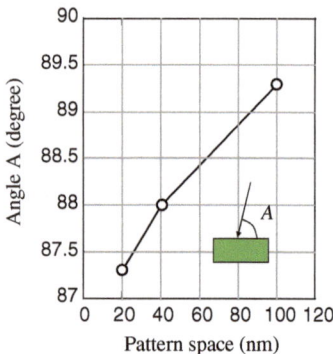

measured with on-wafer sensors were also used (surface potential: -35 V, electron temperature: 4.0 eV, electron density: 5×10^9 cm^{-3}, and sidewall resistance: 3×10^{15} Ω). The ion trajectory at an aspect ratio of six went almost straight to the bottoms of the holes. However, the ion trajectory was distorted at high aspect ratios. Figure 3.13 summarizes the incident angles as a function of aspect ratios. The ion trajectory was distorted at higher aspect ratios.

Figure 3.14 shows the mechanism responsible for distorting ion trajectories in this study. Positive charges were accumulated regardless of pattern space. However, the ion trajectories were not only affected by the positive charges of the bottoms of holes but also by the positive charges of adjacent bottoms of holes. As

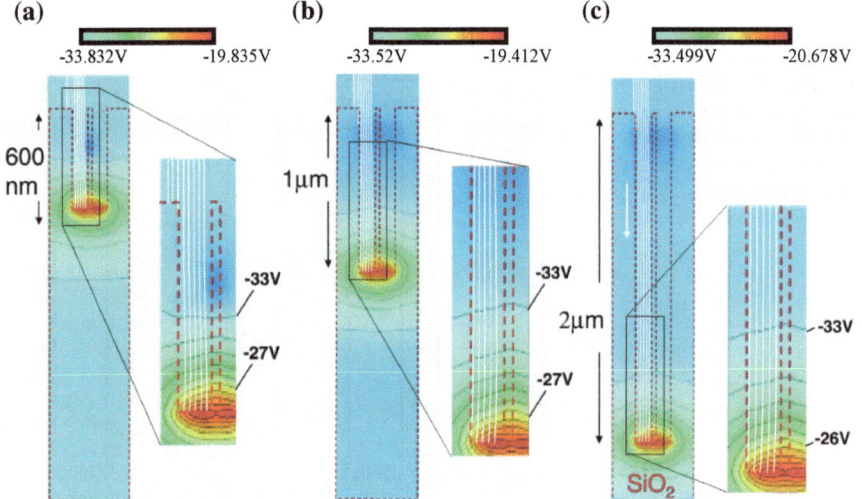

Fig. 3.12 Calculated potential distribution and ion trajectory (*white line*) at aspect ratios of (**a**) 6, (**b**) 10, and (**c**) 20 for pattern space of 20 nm (surface potential: –35 V, electron temperature: 4.0 eV, and electron density: 4×10^9 cm^{-3}, and sidewall resistance: 3×10^{15} Ω)

Fig. 3.13 Incident angle of ions to bottom of hole under conditions in Fig. 3.12

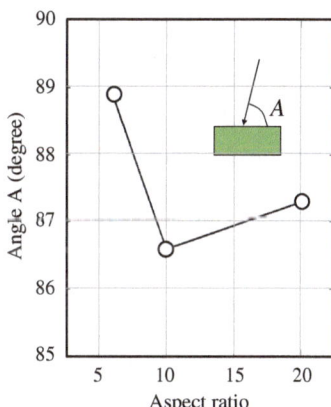

pattern space decreased due to the resists deforming, it is possible that the positive charges that accumulated in adjacent holes affected the ion trajectories. It should be noted that these predictions in Figs. 3.10, 3.11, 3.12, 3.13 do not precisely correspond to the actual distortion of ion trajectories because ion trajectories change with reflections in the shoulders of resists, changes in sidewall resistance, the effects of numerous holes, and other factors [19]. However, pattern space is considered to be one of the main reasons for the generation of twisting profiles, according to these predictions.

Fig. 3.14 Mechanism for
generation of twisting
profiles. As pattern space
decreases due to resist
deformation from (**b**) to (**a**),
positive charges that
accumulate in adjacent hole
affect ion trajectory

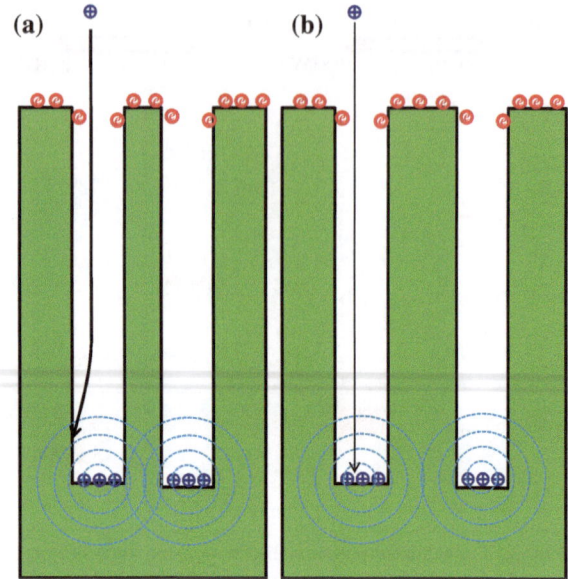

3.4 Conclusions

An ion-trajectory predictive system for use in high-aspect-ratio hole etching was
developed by combining the technique of on-wafer monitoring and sheath modeling. This system revealed that sidewall conductivity strongly affects charge-up
and ion trajectories in high-aspect-ratio holes. It was also revealed that the
decrease in pattern space is one of the reasons for the generation of twisting
profiles. This predictive system is an effective tool for developing nanoscale
fabrication processes.

Acknowledgments We would like to thank Messrs. I. Kurachi, J. Hashimoto, and S. Kawada for
preparing the on-wafer sensors and for the helpful discussions they had with us. We also wish to
thank Mr. T. Ozaki for his assistance with the experiments.

References

1. A.A. Ayon, S. Nagle, L. Frechette, A. Epstein, M.A. Schmidt, J. Vac. Sci. Technol. B **18**,
   1412 (2000)
2. J. Saussac, J. Margot, M. Chaker, J. Vac. Sci. Technol. A **27**, 130 (2009)
3. K. Hashimoto, Jpn. J. Appl. Phys. **32**, 6109 (1993)
4. A. Sankaran, M.J. Kushner, J. Vac. Sci. Technol. A **22**, 1242 (2004)
5. F. Hamaoka, T. Yagisawa, T. Makabe, J. Phys. D **42**, 075201 (2009)
6. Y. Osano, M. Mori, N. Itabashi, K. Takahashi, K. Eriguchi, K. Ono, Jpn. J. Appl. Phys. **45**,
   8157 (2006)

7. Y. Osano, K. Ono, J. Vac. Sci. Technol. B **26**, 1425 (2008)
8. T.G. Madziwa-Nussinov, D. Arnush, F.F. Chen, IEEE Trans. Plasma Sci. **35**, 1388 (2007)
9. J. Matsui, K. Maeshige, T. Makabe, J. Phys. D **34**, 2950 (2001)
10. J. Matsui, N. Nakano, Z.L. Petrovic, T. Makabe, Appl. Phys. Lett. **78**, 883 (2001)
11. K. Nishikawa, H. Ootera, S. Tomohisa, T. Oomori, Thin Solid Films **374**, 190 (2000)
12. G.S. Hwang, K.P. Giapis, Appl. Phys. Lett. **71**, 458 (1997)
13. H. Ohtake, B. Jinnai, Y. Suzuki, S. Soda, T. Shimmura, S. Samukawa, J. Vac. Sci. Technol. B **25**, 400 (2007)
14. T. Shimmura, Y. Suzuki, S. Soda, S. Samukawa, M. Koyanagi, K. Hane, J. Vac. Sci. Technol. A **22**, 433 (2004)
15. B. Jinnai, T. Orita, M. Konishi, J. Hashimoto, Y. Ichihashi, A. Nishitani, S. Kadomura, H. Ohtake, S. Samukawa, J. Vac. Sci. Technol. B **25**, 1808 (2007)
16. T. Shimmura, S. Soda, S. Samukawa, M. Koyanagi, K. Hane, J. Vac. Sci. Technol. B **20**, 2346 (2002)
17. T. Shimmura, S. Soda, S. Samukawa, M. Koyanagi, K. Hane, J. Vac. Sci. Technol. B **22**, 533 (2004)
18. H. Ohtake, S. Fukuda, B. Jinnai, T. Tatsumi, S. Samukawa, Jpn. J. Appl. Phys. **49**, 04DB14 98 (2010)
19. M.A. Vyvoda, M. Li, D.B. Graves, H. Lee, M.V. Malyshev, F.P. Klemens, J.T.C. Lee, V.M. Donnelly, J. Vac. Sci. Technol. B **18**, 820 (2000)
20. S.C. Park, S. Lim, C.H. Shin, G.J. Min, C.J. Kang, H.K. Cho, J.T. Moon, Dry Process Int. Symp. **2005**, 5 (2005)
21. S.K. Lee, M.S. Lee, K.S. Shin, Y.C. Kim, J.H. Sun, T.W. Jung, D.D. Lee, G.S. Lee, S.C. Moon, J.W. Kim, Dry Process Int. Symp. **2005**, 3 (2005)
22. H. Mochiki, K. Yatsuda, S. Okamoto, F. Inoue, AVS 56th international symposium, PS2-MN-WeA-4, 2009

Chapter 4
Feature Profile Evolution in Plasma Processing Using Wireless On-wafer Monitoring System

Abstract Etching profile anomalies occur around large-scale 3-dimensional (3D) structures due to distortion in the ion sheath and ion trajectories. To solve this problem, a system to predict such etching anomalies was developed by combining on-wafer sheath shaped sensor and simulations based on a neural network and a database. The sensor could measure the sheath voltage and saturation ion current density and sheath thickness can be calculated from them. A database was built by using the results from sensor measurements and etching experiments with samples with large vertical steps, which enables prediction of etching shape anomalies from measured parameters. Finally, the system could predict etching shape anomalies around large vertical steps.

Keywords On-wafer sheath shape sensor · Large scaled 3D structure · Ion sheath · Ion trajectory

4.1 Introduction

Precise plasma processes are critical in the fabrication of ULSI and MEMS devices. However, plasma induces damage in devices due to the irradiation of high energy ultraviolet (UV) photons [1, 2] and charged particles [3, 4]. As seen in Fig. 4.1, high energy UV photons from plasma chemical bonds generate defects and degrade the performance of devices. Charged particles cause charge-up damage and anomalies in etching shapes.

We have been developing a system of on-wafer monitoring to solve these problems by combining the results from measurements and simulations [5, 6]. The most remarkable feature of this system is that the sensors to measure plasma irradiation damage can be fabricated using standard microfabrication technology on silicon wafers. The measurements are carried out on the sample stage position of the plasma chamber. Real-time measurements can be accomplished in combination with measurement circuits. Measured data are stored in memory in the

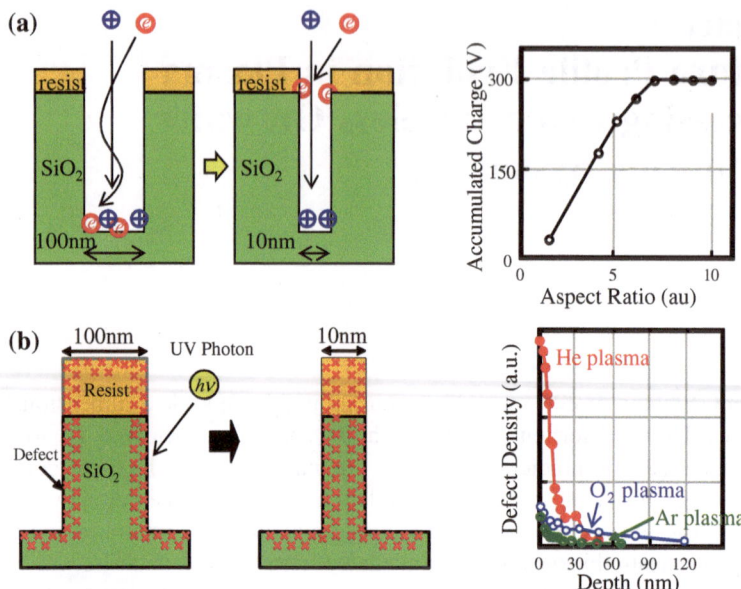

Fig. 4.1 Schematic of plasma irradiation damage: **a** charge-up damage and **b** UV irradiation damage

circuit, and then the measured data can be transferred to a PC through infra-red communication after the sensor and the circuit are unloaded. It is therefore possible to predict damage distributions and etching profiles by combining the results from measurements and simulations [7, 8].

Some MEMS devices have larger scaled 3D structures comparable to the ion sheath thickness on the surface in plasma processing. Because the sheath shape is distorted in such cases due to the MEMS structure, ion trajectories are distorted to the surface and this causes the anomalies in etched shapes shown in Fig. 4.2. We developed a system of on-wafer monitoring to measure ion sheath conditions and predict anomalies in etching shapes to solve this problem.

4.2 Experimental

Anomalies in etching profiles can be predicted by combining the results from sheath measurements and simulations. Etching profiles are mainly ruled by ion trajectories in ion-assisted etching and the trajectories are determined from the sheath electric field. Therefore, we developed a sensor to measure the thickness and voltage of the ion sheath. Figure 4.3 shows the structure of the on-wafer sheath shaped sensor we developed. It has a numerous small electrodes to measure

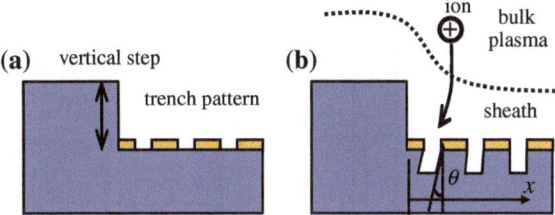

Fig. 4.2 Schematic of etching shape anomaly in 3D structure etching due to ion sheath distortion. **a** Structure of 3D sample. **b** Schematic of distortion in ion sheath, ion trajectory, and etching profile due to step

the surface potential and ion saturation current at the wafer surface. Sheath thickness can be calculated based on the measured results.

The relationship between sheath conditions and etching profiles was investigated with the sensor. Anomalies in etching shapes were investigated by silicon etching using chlorine inductively coupled plasma. Samples with vertical steps and trench patterns were used in the experiment as 3D structured samples, as shown in Fig. 4.2. Ion sheath conditions were measured under the same plasma conditions as those in the etching experiment. As a result, a database of the relationships between ion sheath conditions, 3D step heights, and etching anomalies was constructed.

The system to predict etching profiles was developed based on a neural network. The database was used to learn the neural network.

4.3 Results and Discussion

Figure 4.4 shows the results obtained from measurements with the on-wafer sheath shaped sensor. We found that the sheath thickness and voltage could successfully be measured using the newly developed sensor.

Figure 4.5 has SEM images of 3D samples after chlorine plasma etching. The sidewall of the etching profile was found to be distorted and the distortion was more significant near the vertical step. The closer the distance from the step, the larger the distortion angle. This indicates that distortion was due to distortion in the ion sheath around the vertical step.

Figure 4.6 shows an example of prediction of the etching profile. We found that distortion in the etching profile around the vertical step could successfully be predicted by the predictive system developed by combining the on-wafer sheath shaped sensor and neural network.

Fig. 4.3 Structure of *sheath shaped* sensor

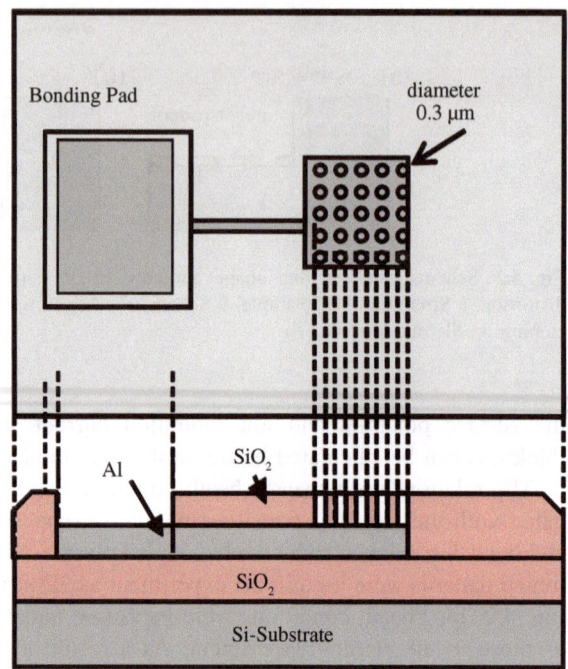

Fig. 4.4 Results from *sheath shaped* sensor measurements of chlorine inductively coupled plasma as function of bias power

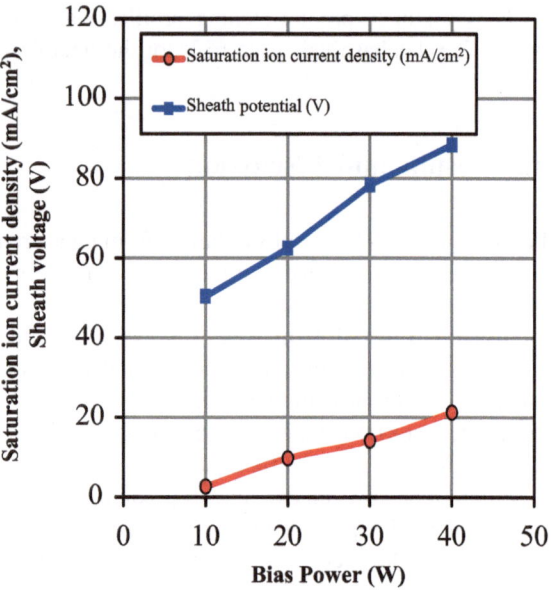

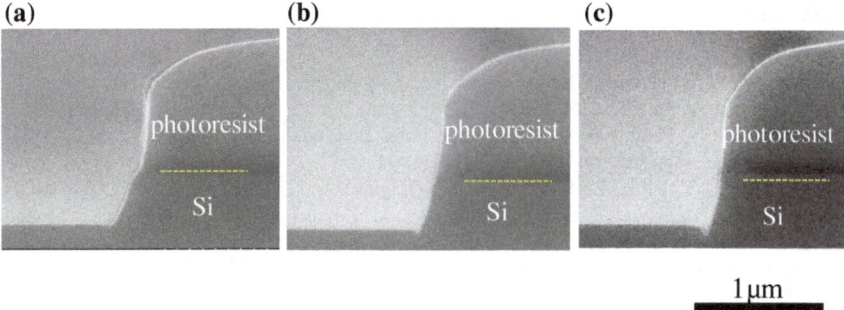

Fig. 4.5 Cross-sectional SEM images of 3D samples after etching at distances from vertical step of **a** 200 μm, **b** 600 μm, and **c** 1800 μm

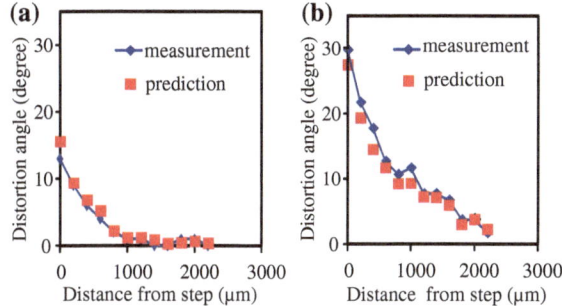

Fig. 4.6 Measured and predicted distortion angles of etching profiles of 3D samples as function of distance from vertical step using mixture gas plasma of Cl_2 and Ar with mix ratio of **a** Cl_2:Ar = 50:50 and **b** Cl_2:Ar = 10:90

4.4 Conclusions

Etching profile anomalies occur around large-scale 3D structures due to distortion in the ion sheath and ion trajectories. A system to predict such etching anomalies was developed by combining the results from an on-wafer sheath shaped sensor and simulations based on a neural network and a database. The sensor could measure the sheath's voltage and thickness. The database was built by using the results from sensor measurements and an etching experiment with samples with large vertical steps. Finally, the system could predict etching shape anomalies around large vertical steps.

Acknowledgments We would like to thank OKI Semiconductor Miyagi Co., Ltd. for fabricating the on-wafer UV sensors. We are also grateful to Mr. Yukihiro Morimoto of Ushio Inc., Dr. Eric A. Hudson of Lam Research Corp., and Mr. Hirokazu Ueda and Dr. Toshihisa Nozawa of Tokyo Electron Technology Development Institute, Inc. for the productive discussions they had with us.

References

1. S. Samukawa, Y. Ishikawa, M. Okigawa, Jpn. J. Appl. Phys. **40**, L1346 (2001)
2. T. Yunogami, T. Mizutani, K. Suzuki, S. Nihimatsu, Jpn. J. Appl. Phys. **28**, 2172 (1989)
3. S. Murakawa, S. Fang, J.P. McVittie, Appl. Phys. Lett. **64**, 1558 (1994)
4. K. Hashimoto, Jpn. J. Appl. Phys. **33**, 6013 (1994)
5. S. Samukawa, Y. Ishikawa, S. Kumagai, M. Okigawa, Jpn. J. Appl. Phys. **40**, L1346 (2001)
6. T. Shimmura, Y. Suzuki, S. Soda, S. Samukawa, M. Koyanagi, K. Hane, J. Vac. Sci. Technol. A **22**, 433 (2004)
7. B. Jinnai, S. Fukuda, H. Ohtake, S. Samukawa, J. Appl. Phys. **107**, 043302 (2010)
8. H. Ohtake, S. Fukuda, B. Jinnai, T. Tatsumi, S. Samukawa, Jpn. J. Appl. Phys. **49**, 04DB14 (2010)

Index

S. Samukawa, *Feature Profile Evolution in Plasma Processing Using On-wafer
Monitoring System*, SpringerBriefs in Applied Sciences and Technology,
DOI: 10.1007/978-4-431-54795-2, © The Author(s) 2014